Global Warming

A Layman's Guide to the Issues

Patrick R. Dugan

AuthorHouse™
1663 Liberty Drive, Suite 200
Bloomington, IN 47403
www.authorhouse.com
Phone: 1-800-839-8640

©2008 Patrick R. Dugan. All rights reserved.

No part of this book may be reproduced, stored in a retrieval system, or transmitted by any means without the written permission of the author.

First published by AuthorHouse   8/25/2008

ISBN: 978-1-4343-9810-9 (sc)

Printed in the United States of America
Bloomington, Indiana

This book is printed on acid-free paper.

Contents

Introduction	vii
Energy	1
Heat	2
Contributions To Global Warming	5
The "Greenhouse Effect"	*5*
"Greenhouse Gas" Is A Misnomer.	*7*
Life Processes and Global Warming	10
Transfer Of Energy By Oxidative and Reductive Reactions And Cycling Of Carbon Dioxide	11
The Role and Significance of Photosynthesis	*12*
Combustion, Forest Fires and Other Wildfires	*13*
Ethyl Alcohol (Ethanol) and CO2 Formation	17
Benign and Beneficial Environmental Effects of Increased Atmospheric CO2.	19
Observations and Opinions	20
Conflictions and Contradictions in the Global Warming Debate.	*20*
It's A Matter Of Choice	*21*
Conclusions	24
On The Lighter Side	26
References and Citations	27
Appendix A	30
Appendix Continued	32

Introduction

Because heat is the only thing that can cause the globe or anything else to increase in temperature; *Global Warming* is a topic primarily concerned with heat. In comparison, *Energy Sufficiency or Energy Security* are topics involved with sources, supply, location, use and cost of all forms of energy. The two topics are tangentially related but are often discussed as a single issue, which leads to confusion and erroneous conclusions. Therefore, global warming should be considered separately from energy sufficiency or energy security and will be discussed separately in this document except where statements are considered from other authors who mix the topics.

Increasing global temperature, therefore necessitates a consideration of heat as a form of energy as well as its sources: sunlight, and fossil fuels as the main reservoirs of stored chemical energy that were ultimately derived from sunlight, as well as existing vegetation, also derived from sunlight. Carbon dioxide (CO_2), a key byproduct formed during the release of energy stored in fossil fuels, vegetation and other organic substances is also considered.

The laws of nature governing all chemical and life processes on the globe, illustrate a reliance on the continuous cycling of carbon and other chemical elements that make up living cells and that the flow of energy in the life processes is unidirectional toward ageing.

Many individuals with an interest in global warming are neither versed in the fundamentals of energy in its various forms nor in the complex chemical reactions in the environment that impinge upon the topics of interest. Consequently, attention is devoted to many of the fundamentals that contribute to our understanding of energy and energy sources, including the significance of thermodynamics, the chemistry and biochemistry of carbon dioxide, photosynthesis and metabolism, and the energy in chemical bonds, which of course hold together all known physical substances.

Although this is a manuscript about biology, chemistry, physics, engineering, economics, and other subjects that impinge on global warming and uses the language of those disciplines extensively; it is not a science manuscript. It is intended for non-scientist professionals with an ability to relate to scientific concepts in simplified terms that they may not have previously contemplated. Where possible, examples have been selected to reflect on conditions and circumstances familiar to most readers. Many energy related technical terms and energy conversion factors are relegated to Appendix (A).

An additional point needs to be clarified at the beginning. That is, not all scientists have identical technical backgrounds within their discipline or even their sub-discipline any more than lawyers, business executives, journalists, marketers, or any other professional groups have the same background within their professional disciplines.

Just as most writing about global warming comes with the built in bias of its authors, this document comes with the perspective of a biochemical process oriented environmental microbiologist--with an attempt to separate technical data and observations from opinions.

Also considered are the larger questions of whether our primary concern resides with: *"Warming the globe"* or *"Energy independence"* or *"The economics of energy supply and use"* or *are there several different constituency groups with different agendas driving separate debates*? Additional questions are considered about the relative importance of *natural* versus *man made* warming.

With an understanding that political decisions control policies already being promulgated about the supposition that carbon dioxide is a primary agent responsible for global warming; the compilation of quantitative technical data, observations and information must speak for itself and leads to the questioning of some of those suppositions. There has been no intention to cover the adequate amount of articles on the purely political aspects of global warming or energy sufficiency (See Ref. 8), other than to say the topics do not appear to be conservative vs. liberal issues; but rather issues brought about by an understandable lack of many biological, chemical and physical, concepts by most policy decision makers.

Energy

Energy is the capacity to do work (activity) and the various types of energy include: chemical, electrical and electromagnetic, kinetic, surface and heat. Energy can be converted amongst its various types and is measured or considered in various ways depending upon the type. Energy may appear, for example, as the kinetic energy of a moving object or as the potential energy possessed by a system because of its position or condition. Hence, a compressed gas or a coiled spring possess potential energy because of its condition; whereas, a brick in the wall of a high rise building has potential energy because of its position. Except for water and wind motion where conversion directly to mechanical energy is possible; most other energy is contained in the form of fuels such as coal, oil, natural gas, wood and other plant products where the energy is stored as internal chemical energy in the bonds that hold the chemicals together. Much of our chemical energy is found in the form of single, double or occasionally triple bonds shared between two or more of the common elements: carbon (C), hydrogen (H), nitrogen (N) and sulfur(S), and in a very different form, phosphate ($PO_4{---}$).

Fossil fuels consist predominantly of a large and complex variety of chemicals comprised mostly of C and H; which is why they contain considerable energy and are flammable or combustible and why many of them are also called hydrocarbons. Each of the above four elements (C, H, N, and S) are also commonly found bonded to other atoms of the same element, e.g. H_2, C_2, C_4, C_6, C_{16}, N_2, S_2, S_8, etc. Carbon has the ability to share four bonds simultaneously with any of several different elements, including some that were not mentioned above. Similarly, nitrogen can share 3, sulfur 2, and hydrogen only 1.

Combustion is a process that releases the stored energy, predominantly as heat and light, usually involving a chemical recombination with oxygen (O); which can be used directly for various purposes such as heating a home. However, to use heat to operate a machine; it must first be converted to kinetic or mechanical energy. The only process where internal chemical energy can be converted directly to kinetic energy is when chemical substances can be combined in an electrolytic cell. Combustion and oxidation will be further considered in a subsequent section.

Table 1 lists some energy sources and some effects of their use.

Heat

Heat is the type of energy arising from the motion of molecules and the greater the motion, the hotter it is. Conversely, the hotter the substance, the greater is its molecular motion and the more chemically reactive it becomes. The globe is warming because the molecules that make up the globe, i.e. earth and its atmosphere are being energized. Solar radiation (Ref. 21) and geothermal, including volcanic activity are the ultimate providers of most of our natural heat energy. The energy found in fossil fuels was derived from once living organisms over geologic time under geological conditions and therefore accrued primarily from solar radiation. During the process of fossilization the easily metabolized components of plant, animal and microbial cells were recycled and the more enzymatically recalcitrant component molecules accumulated to form reservoirs of coal, oil, tar and natural gas. These reservoirs formed as sediments in swamps or at the bottom of lakes and oceans and over time were covered by mineral sedimentary deposits; thereby isolating the reservoirs of fossil fuels from further recycling their contents, until technology provided the means for drilling, mining and other forms of extraction. Thus, fossil fuel reservoirs are often referred to as energy sinks; where the energy is removed from circulation.

Recall that solar radiation is electromagnetic radiation that emanates from the sun and enters the globe; therefore, contributing to warming the globe. Overhead solar radiation on a clear day has a reported average intensity of 1370 watts per square meter at the top of the atmosphere. The annual average reaching the earth surface is reported to be only about 170 watts per square meter; indicating that most solar radiation is either reflected back to space or absorbed as it enters, prior to reaching the earth's surface. We will return to solar radiation in a subsequent section.

Heat is measured in *calories*, defined as the amount of heat energy required to raise the temperature of 1 gram of water by 1 degree C., which is also known as the *small calorie* (cal.). Because larger numbers of calories are frequently used; *1000 calories* or *kilogram-calorie* is often used and designated as: *Cal. or kg.-cal. or kilocalorie or large calorie*. Larger amounts of heat are also frequently measured as *British Thermal Units (BTU)*, defined as the amount of heat energy required to raise the temperature of 1 pound of water 1 degree F. One BTU is equal to 252 calories.

A fundamental thermodynamic property of heat in relation to other types of energy is the tendency of all other forms to change to heat. It is not possible to convert any of the energy types completely into work. For example, some of the energy released in a chemical reaction may be converted to work but some may be released as heat to the surroundings and not convertible to work. That is, when converting energy from one type to another; it never converts at 100% efficiency and some heat is always lost.

This has implications when converting energy released during fossil fuel combustion to generate electricity, followed by transportation of electricity over wires to locations remote from a power generating plant where it might then be used for various purposes such as charging batteries (i.e. converted to chemical energy). A considerable amount of energy is lost as heat, both during transportation of electricity over wires and again during the conversion of electricity to chemical energy; thereby decreasing the overall efficiency of using fossil fuels to charge batteries on a large scale. One 2002 estimate of the average heat rate i.e. BTU content of fuel burned, divided by net kilowatt hours (KWH) of electricity generated was 45% to 50%.

The Laws of Thermodynamics, which describe fundamental energy relationships, are presented in more detail in Appendix (B). Beyond that, the reader is referred to any college level text on Physical Chemistry, or Thermodynamics published in the last 50 years.

However, we need to be aware that to legislate contrary to these laws of nature is a fools errand. No one can lower the global temperature by merely converting to alternate forms or sources of energy to accomplish the same or greater amounts of work because all the processes produce heat and some heat is always lost in every process. Further; although there is the equivalent of 778 ft. lb. of work in 1 BTU; no one can obtain 778 ft. lb. of work from a BTU of heat in any form of engine because there is always heat lost as exhaust and 100% conversion of energy from any fuel is not possible in any type of engine. On occasion, someone or some group believes they have discovered a process where they obtain more energy from a system than was put in and it creates a temporary flurry of great interest until others cannot duplicate the work. The last such episode of this type occurred in 1990 when *cold nuclear fusion* was reported.

Most large scale energy expenditure is used to generate heat that is used directly for a variety of purposes including: heating homes and other buildings, or for heating water used for cleaning, cooking, sterilizing, etc. Liquid water is also converted to steam by heating and the expanded hot water vapor is used to turn crankshafts, rotors, turbines, wheels, etc. in steam engines. For example, energy may be converted in turbines to electricity, or in internal combustion engines to drive pistons and crankshafts where the energy can ultimately be converted to motion. In every case a considerable amount of the heat is lost as exhaust to the environment.

Cooling by refrigerators and air conditioners is the transfer of heat from a relatively small closed environment to a larger external environment. This is accomplished via the use of energy to compress a coolant gas partially to a liquid contained in a cooling coil, then pumping it to a warm space where heat is transferred from the space to the cooling coil followed by pumping the coolant through a throttling valve or an expansion valve, emerging as a mixture of liquid and vapor at a cooler temperature, then to an external location where the coolant expands and releases heat to the environment; usually with the aid of a radiator and fan. Refrigeration units and air conditioners are net generators of heat.

All cooling towers, smoke stacks, chimneys, exhaust pipes and appliance vents are conduits designed to transfer excess heat in the form of hot gases, vapors or solid micro particulates directly to the atmosphere.

Large industrial processes often use cooling water to remove heat from process streams. Power plants are frequently located on a sea or lake or river shore so the cooling water that becomes heated can be discharged directly into a large body of receiving water. This, of course, raises the temperature of the receiving water and can have biological effects. For example; Manatees migrate up the Caloosahatchee and other rivers from the Gulf of Mexico seeking the warmer water near an electric power plant.

In locations where cooling water must be reused, the water must first be re-cooled. Cooling towers are used to transfer the heat absorbed by the cooling water to the atmosphere by allowing the heated water to cascade over and through slats or grids designed to increase the contact surface area in the tower and allow maximum exposure of heated water to cooler air. Photographs of nuclear as well as coal fired power plants frequently show cooling towers or stacks with clouds of water vapor rising into the atmosphere. It is interesting to observe that many news articles, including those on television, while using such photographs as illustrations of global warming, seem to fixate on CO_2 as the source of warming rather than on the primary heat source; which in this case is the evaporation of hot water vapor mixed with air from a cooling tower. The fact that it is rising into the atmosphere indicates that the air-vapor mixture is warmer and therefore less dense than the atmosphere.

Table 1. Some Energy Sources And Some Effects Of Their Use

Fuel or Alternate Energy Source	Reduction of Dependence On Foreign Oil ?	Subsidize Farmers?	Decrease Atmospheric CO2 ?	Decrease Global Warming?	
Coal	Yes	No	No	No	See Footnote 1
Imported Oil	No	No	No	No	
Natural Gas	Yes	No	No	No	
Bio-methane	Yes	Possibly	No	No	
Bio-ethanol	Yes	Yes	No	No	
Other Bio-fuels 1	Yes	Possibly	No	No	
Geothermal	Yes	No	Yes	No	See Footnote 1
Hydrogen	Yes	No	Yes	No	
Hydro-electric	Yes	No	Yes	No	
Solar	Yes	No	Yes	No	
Wind	Yes	No	Yes	No	
Tidal change	Yes	No	Yes	No	
Nuclear	Yes	No	Yes	No	See Footnote 2

Footnote 1. It takes the same amount of energy to accomplish a given amount of work regardless of the source. However, the cost of different sources may vary considerably depending on many factors such as safety, maintenance, transportation etc.

Footnote 2. Nuclear energy involves changes in the nucleus of an atom and amounts of energy and heat released are several orders of magnitude greater than for the chemical reactions or chemical bonds discussed in this document.

Contributions To Global Warming

Discussions about global warming frequently appear to deteriorate to an "all or none" argument about natural vs. man made sources. The real questions should be: *What are the individual sources of heat and what is their relative contribution to the total? How much heat originates from each source and what can or should be done about each heat source? Do we really want to curb global warming on a grand scale and what would be its implications?*

In this context it is important to recall that small individual contributions from very large numbers of sources can often contribute as much or more than contributions from a few very large heat sources; a consideration that is reminiscent of the instructive grade school ditty: "Little drops of water and little grains of sand make the mighty oceans and the pleasant land".

Actually, the trapping of any heat from a source external to the globe, (the sun), as well as the production of any amount of heat, however small, from any source in the world over any time period is a contributor to global warming. When one thinks about it; this means that striking a match, or lighting a candle, cigarette, propane grill, campfire or running a motor, engine, or electrical device of any kind, all contribute to global warming. The distinction between *contribution to* vs. *cause of* also needs to be emphasized.

The "Greenhouse Effect"

There is little debate as to whether the earth is in a warming trend. Global temperature appears to have increased by about one degree C. over the past 120 years. (See Appendix C, possible Krakatoa temperature effect). Simultaneously, the amount of atmospheric carbon dioxide (CO_2) has increased from about 280 milligrams per million parts (ppm) of air to about 360 ppm. Many articles on the subject make a giant leap to the conclusion that CO_2 is the primary cause of global warming, based upon (A) the observed correlation of increased CO_2 in the atmosphere and the global warming trend and (B) the premise that atmospheric CO_2 traps heat caused by solar radiation at or near the earth surface, preventing it from reflecting back to outer space; thereby causing the global temperature to rise. This was thought to be akin to the heating effect observed when sunlight penetrates the glass of a greenhouse and interacts with the benches, soil and other objects in the greenhouse; whereupon some of the electromagnetic radiation of sunlight is converted to heat that is trapped in the greenhouse. It is often referred to as the "Greenhouse Effect" and CO_2 is being frequently referred to as a "greenhouse gas".

Approximately 70% of incident solar radiation is absorbed by clouds, gases, vapors and solid particulates in the atmosphere, such as oxygen, ozone, water vapor and smoke and by the earth's surface. The earth and its atmosphere also radiate infrared (IR) wavelengths in all directions with some of it radiating back toward space. Some of this is also absorbed by substances in the atmosphere thereby preventing outgoing IR radiation from re-entering space. Trace gases, vapors and substances that strongly absorb IR wavelengths are called relatively important trace species (RITS). Because RITS partially trap outgoing IR they somewhat mimic the heat trapping effect of glass in a greenhouse and are extensively referred to as "greenhouse gases". The role of RITS in altering the radiative energy balance is called radiative forcing. The value depends on absorbing characteristics of the molecules and their calculated lifetime as well as their concentration in the atmosphere. Comparison of radiative forcing tendency among various gases and vapors is accomplished by computing the changes in radiative energy flux or potential surface temperature due to a projected atmospheric concentration of each gas or vapor. Table 2 shows relative radiative forcing potential for a few RITS.(Refs.37,38). One has to question what is precisely meant by the calculated lifetime of CO_2 in the atmosphere in lieu of the magnitude of the annual oscillations for CO_2 reported from Mauna Loa, Hawaii. The annual oscillation of CO_2 during the period 1957 to 1989 is explained by the annual cycle of photosynthesis and respiration of plants in the Northern Hemisphere, resulting in a remarkably consistent yearly decrease of about 6.3 ppm CO_2 with an average yearly deviation of only about 0.4 ppm. The comparable yearly average increase is about 7.3 ppm with an average deviation of 1.8 ppm. This implies there is almost no change in photosynthesis and plant respiration over the 32 year interval and the net annual increase in atmospheric CO_2 is about 0.92 ppm with considerably greater annual deviation. However, the variations do not appear to correlate well with temperature records in the Northern Hemisphere.

Table 2. Relative radiative forcing potential for some trace RITS. From (Ref.37)

Trace Gas	Radiative Forcing Multiplier Relative To CO_2
Carbon dioxide (CO_2)	1
Methane (CH_4)	21
Nitrous oxide (N_2O)	206
CFC-11 (Freon-F 11; $CFCl_3$)	12400
CFC-12 (Freon-F 12; CF_2Cl_2)	15800
Halon 1301 (CF_3Br)	16000

The conclusion that CO_2 is a primary factor responsible for the observed global warming trend; based on very little quantitative data, is challenged in this document. Statisticians as well as most scientists and engineers are aware that correlations are not necessarily cause and effect relationships and there is scant quantitative data to show a cause and effect relationship between the amount of CO_2 in the atmosphere and the amount of observed temperature change. *Proponents of the greenhouse gas supposition need to provide their quantitative data showing the relationship between amount of heat retention and concentration of CO_2. Further, correlations between global temperatures and water usage; particularly irrigation, construction of dams and other water impoundments, the use of cooling towers and moisture released from wildfires need to be examined.*

As to whether the observed warming trend is natural or man made; simple observation shows that global warming is a net effect of natural causes and hundreds, probably thousands, of man made contributions. In addition to CO_2, the observed global warming also correlates with hundreds of processes; including growth in world population, the production, harvest and use of food sources, as well as the development and use of electricity to name a few. With tongue in cheek we can't resist pointing out the correlation between global warming and the production of such things as:

clocks, shoes, eating utensils and golf balls (perhaps balls in general) over the past 100 to 150 years. Ridiculous? Perhaps; but it does illustrate the point that correlations are not necessarily causes.

The more important question to be answered is: What is the relative amount of warming due to man made causes in comparison to natural causes over which there is relatively little control?

"Greenhouse Gas" Is A Misnomer.

There is such a thing as a "Greenhouse Effect", meaning warming in a space enclosed by glass where light has penetrated. However, there is no such thing as a natural "Greenhouse Gas". Air is the gas in the greenhouse and it normally contains a small amount; approximately 360 ppm, of CO_2, which is not responsible for the heat generated in a greenhouse. Recall that water vapor and moisture in general is much more prevalent in air and this amount of moisture absorbs infra-red and near infrared wavelengths much more extensively than any of the other constituents of normal air. Also recall that vapors are not gases. The heat generated when sunlight passes through glass and impinges on solid material components such as plants, benches or moist soil in a greenhouse is not primarily due to CO_2. Neither is CO_2 the primary cause of heat generated in an automobile when it is parked in the summer sun with its windows closed.

Although it is true that CO_2 can contribute to the conversion of solar light to heat; so will any multiple bond molecule found in the light path. Any gas, or vapor such as solvent vapors and moisture, or solid particles such as dust and smoke or any of thousands of other volatile substances that find their way into the atmosphere will absorb some wavelengths of solar radiation and become energized and in a similar manner will also contribute to global warming more effectively than will CO_2. This phenomenon is the basis of all ultraviolet, visible and infrared spectrophotometry; a well known analytic technique where specific chemicals can be identified by their absorption of specific wavelengths of electromagnetic radiation, and which are also associated with discreet energy levels. Extensive listings of absorption spectra of individual chemicals and specific chemical bonds are readily available in handbooks. (See Ref. 5, CRC Handbook of Chemistry and Physics). This phenomenon is also the basis for heating and cooking food in microwave ovens and in many other applications. Readers may be aware that any aroma, odor, fume or stench in the air that can be detected by smell; whether from freshly brewed coffee, a rose garden, new mown hay, decaying fish on a beach or sewage indicates there are molecules in the atmosphere that are capable of absorbing infrared light and converting it to heat.

Frequent meteorological forecasts during televised weather reports that, "a cloud blanket on a cold night will retard heat loss to outer space and keep the over night temperature from dropping as much as would be expected on a clear night" is also a well known analogy to what is being called the greenhouse effect. Clouds are water vapor-not CO_2 and the amount of atmospheric moisture in most locations far exceeds the amount of CO_2 in its ability to absorb solar energy. Atmospheric moisture content at a specific location frequently exceeds 2%; whereas the average CO_2 content is only about 0.00036 %. Figure 1 is a NASA photograph taken from outer space and clearly shows some of the extensive cloud cover around the globe.

One merely has to place their hand on a lighted electric light bulb to recall that lights are a significant source of heat. A 100 watt incandescent bulb generates about 341 BTUs of heat per hour. During the 1960s and 1970s building designers often planned to utilize the heat generated by lights in a building as a supplemental source of heat for the building. In this situation, curtailing the use of incandescent lights may also require supplemental replacement heating for a particular building.

Figure 1. Photograph of earth from space showing extensive cloud formations around the globe. From Image Science and Analysis Laboratory, NASA-Johnson Space Center, Appolo 17.

Figure 2. Composite Photograph of the Earth's City Lights. The brightest lighted areas are the most urbanized around the globe. From U.S. Defense Meteorological Program (DMSP) Operational Linescan System. Image by Craig Mayhew and Robert Simon, NASA, GSFC based on DMSP data.

Figure 2 is a composite of the earth's city lights rendered by NASA and DMSP; illustrating that the worlds population is increasingly lighting the dark side of the world. The brightest lighted areas are the most urbanized areas around the globe, and therefore relate to locations of world economic development.

It is recognized that CO2 "domes" often build up over cities due to concentrated use of fossil fuels and localized climatic conditions. George, et al reported that the atmospheric CO2 content and temperature over the center of Baltimore Md. was elevated by an average 66 ppm over 5 years (14.8 C.) compared to a suburb (13.6 C.) and a rural site (12.7 C.) along a 5 mile transect. (See Ref. 12). However, other investigators compiled data showing that there is actual cooling in various cities over time. "CO2 Science" reports each week on the temperature record from one of the 1221 United States Historical Climatology Network stations for the period 1930 to 2007 and concludes that during this period of rapid CO2 atmospheric buildup there is no significant increase in average temperature (Ref. 2, 4.).

Also well known and documented is that stone, concrete, asphalt roofing and other structural building and paving materials absorb sunlight and convert it to heat and retain and re-emit heat for a considerable time after the sun sets. This is also a reason why cities often tend to be warmer than their surrounding suburbs. Those who have fireplaces in their homes are aware that the bricks or stone continue to emit heat long after the fire is out.

In addition to moisture, other vapors as well as dust, smoke and other solid micro-particulate materials, and gases such as: methane, ethane, propane, ozone, some oxides of nitrogen and sulfur, etc. are known to contribute much greater absorption of solar energy and conversion to heat than does CO2.

Color is imparted in organic substances by the selective absorption of specific wavelengths of light and lack of absorption of other wavelengths. It is primarily due to the configuration of alternate single and double shared (covalent) chemical bonds, called conjugated double bonds, and is a familiar factor in the conversion of solar radiation to heat in colored materials. For example; if one places their hand on the hood of a dark colored car parked in the summer sun, it is much hotter than a white or light colored car parked next to it. Or if you were to walk barefoot in the grass on a bright sunny day and then step onto a blacktop driveway, or witness snow melting over the driveway in the winter sun compared to over the grass; you would notice the heat difference.

None of this type of heat generation is attributable in significant measure to CO2 trapping of solar radiation via a greenhouse effect. If it was; then other methods of trapping solar energy, such as with the use of solar panels, that convert light to electricity or heat would also be significant contributors to global warming.

An additional factor related to color is the reflective capacity of snow and ice. Light absorption by dark colored pavements has already been mentioned but we are spending considerable effort to rid snow and ice from an expanding area of roadways, sidewalks and parking spaces in the winter; which decreases snow reflectivity and exposes darker colored components of the earth, thus accommodating the conversion of sunlight to heat. The relationship of light and color will be further discussed in the subsequent section on the role of photosynthesis.

Life Processes and Global Warming

Life processes also contribute to global warming. Nearly everyone is aware that healthy humans maintain a relatively constant temperature around the clock of about 98.6 degrees F (37 degrees C.). The amount of heat constantly produced by the average normal human body "at rest" is calculated as 400 BTUs per hour (see Appendix D). Architects and building design engineers take this heat into account when designing buildings that hold large numbers of people such as theaters, schools, auditoria, convention centers, etc. In effect each human body as well as each warm-blooded animal is equivalent to a mobile heating device when the ambient temperature is less than body temperature. As the population of humans, presently estimated at about 6.7 billion and the number of animals increase, the amount of heat released becomes significant. It is calculated that about 23 quadrillion BTUs (23 QUADs) of heat are released annually from humans at rest and the amount increases with work, exercise and other activity in addition to the incalculable amount from animals.

A fascinating implication of this is that all health activity aimed at losing weight by "burning calories" e.g. athletics, exercise programs and strenuous exertion of any kind is in contradiction to the stated "global warming" control objectives of curtailing CO_2 production and release of heat to the environment.

Another familiar biological process is municipal sewage or domestic waste water treatment; a process used to provide the oxygen demanded by microorganisms as they metabolically oxidize the organic waste material found in the sewage. Sanitary engineers refer to this as removal of biochemical oxygen demand (BOD) from the waste water. It is carried out in a treatment facility where sufficient oxygen is supplied to oxidize any dissolved or suspended organic waste via the growth of microbes so that untreated waste water doesn't flow to a receiving stream or lake where it could consume oxygen and suffocate fish and other desirable wildlife. BOD removal is a desirable, most would say an essential, heat and CO_2 generating oxidative process where the heat flows off to a receiving stream. In addition, waste water treatment removes disease causing microbes.

Transfer Of Energy By Oxidative and Reductive Reactions And Cycling Of Carbon Dioxide

Many appear to be unaware of the ubiquity and prevalence of CO2 in the environment; that it is an essential metabolite for all living organisms including the cells of all microbes, plants and animals, and how it cycles among organisms to become incorporated into all organic substances. Because CO2 and carbon are so pervasive in the natural world it would take several volumes to detail all of the atmospheric science, biology, chemistry, geology, and physics involved in its cycling. Rather, there are good overviews (See Ref. 29, Carbon Cycle, NASA) as well as more detailed compilations of data available (See Refs. 11, 38). A few additional paragraphs on CO2 and its transfer among living cells is warranted.

At ambient temperatures, CO2 is a gas. However, CO2 can be condensed under pressure with cooling to about -78.5 degrees C. to produce solid CO2; a familiar substance often called "dry-ice" that is frequently added to frozen packaged meat and other frozen perishables to accommodate shipping in the frozen state. Above -78.5 C. solid CO2 sublimes directly to gaseous CO2 without going through a liquid phase and it returns to the atmosphere.

When CO2 gas dissolves in water it associates with the water to become carbonic acid (H2CO3). Its solubility increases when placed under pressure and when added to water under pressure; it becomes the familiar "club soda", "soda water", "sparkling water" or when flavoring and other ingredients are added it is sold as "Coke", "Pepsi" and other soft drinks. Another common use of CO2 gas under pressure is in some types of fire extinguishers. Being heavier than air and non-combustible, it tends to "suffocate" small fires by depriving them of oxygen in the air.

Carbonic acid reacts readily with alkaline solutions, i.e. above pH 7.0, to form common reactants such as sodium carbonate (Na2CO3), and sodium bicarbonate (NaHCO3), a chemical included in all baking powders along with a mild edible acid that reacts with it during leavening or baking to release CO2 gas, which then causes the bread or cakes, etc. to rise. The pH of the world's oceans and most fresh water bodies is slightly alkaline with a pH range of about 7.9 to 8.3, thereby acting as a repository of CO2 as carbonate in association with many metal ions. Calcium carbonate is a major component found in sea shells, bones, limestone, chalk, concrete cement and many other familiar materials; all of which will react with acid to release CO2 back to the environment. As an aside, this type of chemical reaction is likely to have been responsible for much of the deterioration of limestone buildings caused by acid rain over the past 200 years or so.

Convenient or not, all tangible substances in the physical world consist of chemicals formed from natural or manmade molecules of varying complexity. All living or once living fossilized organisms, whether animal, microbial or plant, consist of carbon based organic substances, including the fossil fuels which were formed at various geological times at various locations from various mixes of organisms under different environmental conditions.

In certain chemical and biochemical reactions that involve the transfer of electrons, the phenomenon is known as oxidation/reduction (O/R or redox reactions). In the language of chemists *any atom or molecule that loses electrons is said to be oxidized and any molecule that gains electrons is said to be reduced*. Reduction in this context does not mean decreased. The significance to this discussion is that oxidative reactions release energy stored in chemical bonds; whereas, reactions that add electrons to a molecule require energy input to form those chemical bonds. Reduced substances, then, are energy rich and oxidized substances such as CO2 are energy deficient. It is also common that many substances are only partially reduced or partially oxidized.

Most of the discussion in this document is concerned with organic substances, although O/R also pertains to minerals and non-organic chemicals as well. The bottom line relative to carbon in the environment is that CO_2; which is the furthest that carbon can normally be oxidized, is increasing in the atmosphere and therefore the oxidative half of the carbon cycle is out of balance with the reductive half cycle that reduces CO_2 back to organic and other chemicals. Oxidation/reduction can be thought of as somewhat analogous to constructing a high rise building. Raising bricks up several floors and laying them in place requires energy expenditure. When the building remains in place it contains potential energy and when the building is ultimately razed the energy is released.

All animals and many microbes must absorb reduced high energy nutrients, metabolize them in order to release the energy to carry out cellular functions such as motion and growth and then excrete the by-products of metabolism. They require as a nutrient any cellular component they cannot synthesize. That is, they take on oxidizable organic energy sources (food) and release that energy from reduced carbon bonds in the form of electrons, always in association with hydrogen ions, through a complex series of steps (digestion and metabolism), terminating with a final step that transfers the electrons to oxygen, forming water. In this case the carbon is released as CO_2. This process is called *aerobic respiration* and is the reason all animals must breath or otherwise consume oxygen while metabolizing reduced organic nutrients to CO_2.

Caution must be exercised as to the environmental usage and meaning of the words cycle, cycling and recycle. They are often used interchangeably in chemical and physical contexts. For example CO_2 is a component in the O/R scheme of the carbon cycle but carbon containing chemicals also change from a solid or liquid to a gas during the cycling process and can be considered as cycling from a sub-surface reservoir to an atmospheric gas just like we view water cycling in the context of hydrological movement in its various physical states in the environment (e.g. vapor, precipitation, snow, rivers, ocean, transpiration, etc.). Further, carbon in nature is nearly always associated with hydrogen, nitrogen, oxygen, sulfur and other elements and is therefore always a simultaneous component of other elemental cycles. The jargon has become more confusing as the result of the term recycling to mean return, re-manufacture and re-use of a product without reference to either the chemical or physical state of the product. Hence, glass bottles, metal cans and paper or other organic carbon substances such as found in trash and sewage are recycled.

An additional cautionary note related to consideration of elemental cycles and carbon cycling in particular, is that the cycles are merely a matter of convenience to explain what is happening to a particular element. They never occur in isolation from other elements and there is no size dimension placed on the cycles. The amount of carbon in the atmosphere is increasing over time. Therefore, the amount of carbon that needs to be recycled to keep the cycle in balance is also increasing. This means that it will take considerably more forest acreage today to sequester the larger amount of CO_2 found in the atmosphere compared to the amount of forest lost for whatever reason 20, 50 or 100 years ago.

The Role and Significance of Photosynthesis

Going green is a popular term used by marketers to proclaim that most every current activity and issue is related to global warming and the environment. The green, of course, refers to the color of vegatation which is due to the presence of the green plant pigment chlorophyll.

The photosynthetic process found in all green plants and some photosynthetic microbes requires CO_2 which the plant or microbe absorbs from the atmosphere and reduces the CO_2 to form organic plant constituents such as sugar, cellulose, starch, green chlorophyll and other pigments and all other cellular components, using the energy from solar radiation (light) and hydrogen obtained from the photosynthetic splitting of water (H_2O). The reactions are accompanied by the release of oxygen (O_2) to the atmosphere also from the splitting of water. The content of most plants is approximately 80 % water and 20 % solids.

About 60 % of the dry weight solids of trees and other plants including grasses consists of reduced carbon compounds that are therefore energy rich materials.

Sugars such as: fructose, glucose, maltose, and sucrose are found in various plants. Each of these sugars is a molecule that either contains six carbons or two such molecules bonded together. Starch and cellulose are polymers (chains) of glucose molecules.

As mentioned previously in conjunction with the discussion of heat released by humans; all animals and many microbes must use oxygen from the air to oxidize and slowly release the chemical energy found in plant components, and convert the energy to heat, motion, water, CO_2, partially oxidized waste products and synthesis of new cells; i.e. growth. The world human population generates more than 4.4 billion metric tons of CO_2 annually while breathing at rest. Any human (or any animal) activity that increases respiration e.g. work, exercise, and play also increases the amount of CO_2, heat and other by-products released to the environment. At present we have no reasonable means of estimating the global contribution of animals in this context because we do not have reasonable estimates of their global numbers.(again, see Appendix D for calculations).

Molecules that absorb visible light are called pigments and all photosynthetic organisms contain some form of chlorophyll, a green pigment molecule that absorbs specific wavelengths of light and transfers the energy to form other cellular energy molecules; in this case high energy phosphate bonds.

Sunlight contains a range of wavelengths of different energy levels including; gamma rays (0.001 nanometers to 1.0 nm), X-rays (1.0 to 10 nm), ultraviolet light- sometimes called black light(10nm to about 380 nm), light visible to humans (about 380 nm to about 750 nm), near infrared light (about 750 nm to about 1000 nm), infrared (about 1000 nm to 0.01 centimeter), radio waves (longer than 0.01 cm).

The part of the spectrum visible to most humans includes: peaks around 400 nm that appears violet, 450 nm-blue, 550 nm-green, 575 nm-yellow, 630 nm-orange, and 650 to above 700 nm-red with shades in between the wavelength peaks.

Most of the higher energy, shorter wavelengths, of sunlight do not reach the earth surface because they are absorbed by oxygen and ozone in the upper atmosphere and much of the lower energy, longer wavelengths, are absorbed by water vapor in the air.

Each atom or molecule has a characteristic range of wavelengths that it absorbs. Chlorophyll molecules absorb light predominantly in the blue and red portion of the visible spectrum but transmit the green and yellow wavelengths, which is why the pigments are green or yellow-green. Trees and most other photosynthetic organisms filter out red and near infrared wavelengths of light; thereby converting the energy of these "warmer" wavelengths to reduced chemicals, some of which are then used to reduce CO_2 to organic cellular products. This selective filtering effect of trees and forest canopy is also the reason why it is cooler under shade trees than in the direct sun and is another reason why loss of forests allow for an increase of global warming. Some plants also produce yellow or orange pigments such as carotene and other carotenoids as well as red and blue phycobilins that have not been considered here.

Combustion, Forest Fires and Other Wildfires

Combustion of any organic material including: wood, plants, paper, most synthetic plastics, fossil fuels and other petrochemicals, etc. also produces CO_2 and is analogous to the aerobic metabolism of animals except that the oxidation is very rapid, resulting in fire and releases light energy in addition to heat, CO_2, moisture and partially oxidized organics. In the 26 year period between 1980 and 2006 over 2000 structures were destroyed by wildfires in the United States. In addition, over 4 million acres of forest burned annually in the United States. The number of acres burned annually is increasing. Over the 7 year period between 2000 and 2006 over 5 million acres burned per year (Ref. 30). Yet, in 2002 when the federal executive government proposed to reduce future forest fires by thinning the underbrush and forest density, the proposal was rejected by Congress.

Forest fires and wildfires, like house fires, are particularly odious because in addition to the release of heat, moisture and CO2 to the atmosphere, there is an enormous and rapid release of smoke particles and noxious vapors which are serious health hazards, particularly for the breathing impaired. With reference to forest or wildfires it also needs to be emphasized that vegetation destroyed by fire in addition to converting a valuable resource directly to CO2, is no longer capable of photosynthetic removal of CO2, thereby producing a double jeopardy relative to balancing the carbon cycle (Ref. 10). Some wildlife experts argue that wildfires are healthy for the forest environment because re-growth of vegetation will occur and the forest will be renewed in 100 to 150 years. However, in some locations, for example Yellowstone Park, which burned extensively in 1988, it takes over 100 years for a lodge pole pine to reach a 30 foot height. Meanwhile, whatever remains of the dead trees become infested with insects and the insects in turn support the population of birds such as three toed woodpeckers. However, the loss of tree root systems that normally hold soil in place contributes to mud slides on hill slopes. The "trade off" seems to be the loss of 5 million acres of forest per year, compensated by a bumper crop of insects that leads to increased populations of woodpeckers, etc.

Figure 3 is an aerial photograph of a crown fire in Yellowstone Park in 1988 and Figure 4 is a photograph of a ground fire in Yellowstone Park in 1976. Both photos clearly show the extensive generation of smoke and moisture plumes.

Figure 5 is an aerial photograph showing an extensive dust plume due to Santa Ana winds in Baja California in 2007, illustrating an additional source of solid particulate material suspended in the atmosphere that converts sunlight to heat.

Figure 3. Photograph of canopy fire burning in Yellowstone National Park in 1976 showing the extensive generation of smoke and moisture plumes. Courtesy of National Park Service Fire and Aviation Program, National Interagency Fire Center, U.S. Department of Interior.

Figure 4. Photograph of canopy fire burning in Yellowstone National Park in 1988, showing extensive smoke and moisture plume that spread hundreds of miles South into Utah and Idaho. Courtesy of National Park Service Fire and Aviation Program, National Interagency Fire Center, U.S. Department of Interior.

As a very rough approximation; the annual amount of CO_2 produced from the average 5 million acres burned per year in the U.S. during the years 2000 to 2006 was 1.15 billion tons of CO_2 and the average amount of CO_2 sequestration capacity lost was about 1 billion tons per year. In addition, about 1.22 billion tons of moisture was released to the atmosphere, not counting soil moisture lost. These values are presented only as gross estimates to illustrate that there are significant releases to the atmosphere as the result of forest and other wildfires. Calculations are described in Appendix E.

Figure 5. Aerial photograph of dust plumes caused by Santa Ana winds blowing across farmland in Baja California, Mexico, October 2007. The Pacific Ocean can be seen in the upper right corner. Photograph by Image Science and Analysis Laboratory, Johnson Space Center, International Space Station Program, NASA.

Wild fires occur around the world every year. If the total global annual production of new photosynthetic activity in each year does not equal or exceed the total lost due to fires the following year, it results in a net decrease of global photosynthetic CO_2 sequestration capacity over time and will be reflected as an increase in atmospheric CO_2.

Ethyl Alcohol (Ethanol) and CO2 Formation

"Tiny Bubbles In The Wine"; a song made popular by the Hawaiian singer Don Ho reminds us that the gas bubbles in the wine and other fermented beverages is CO2. Anyone who has ever produced wine, sparkling wine, beer, champagne, hard cider, "white lightning" or other alcoholic beverages is aware that the bubbles emanating from the brew are CO2 gas. A question then arises: If CO2 is legally defined as a pollutant, does it mean that our commercial baked goods or carbonated beverages of all kinds are polluted? Does it mean that anyone who opens a container with a CO2 content is a polluter?

In alcoholic beverages where the CO2 originates from the conversion of sugar, starch, or cellulose through the metabolic action of yeast cells and other microbes; the process is called fermentation. The ferment first converts starch or cellulose polymers and disaccharide sugars that each contain 12 carbon atoms to their simple component sugars, called hexose monosaccharides that each contain 6 carbon atoms. It then further converts each 6 carbon sugar, through a series of biochemical reactions, to two molecules of ethanol (CH_3CH_2OH) plus two molecules of CO2 gas; regardless of whether the sugar or starch was originally derived from fruits, sugar cane, beets, tubers e.g. potato or any of several different grains or cellulose from any of several different grass or tree species. Therefore, for every ethyl alcohol molecule produced by fermentation, a molecule of CO2 has also already been produced. Separation of ethanol from the liquid ferment followed by subsequent packaging and shipping requires additional energy expenditure with its extant production of CO2.

Many recent articles tout the use of ethanol as an alternative to gasoline or diesel fuel for automotive use on the grounds that it is cleaner burning and produces less "greenhouse gas"; presumably meaning CO2. However, the comparison is flawed, because the source of ethanol is starch or sugar or cellulose and CO2 is produced both during manufacture and again when it is burned. Ethanol, is a partially oxidized liquid that has already yielded 44 tons of CO2 for every 46 tons of ethanol produced during its manufacture. Therefore, production of 10 billion gallons of ethanol (density = 0.79 at 20 C.) will have released about 39 million tons of CO2. Then, when burned as a fuel, the 10 billion gallons of ethanol yields an additional 39 million tons of CO2.

Added to the above burden is the amount of CO2 produced during production of beverage alcohol. In the United States during the year 2002, it was reported that 153 million 9 L. cases of spirits were sold. Assuming the spirits were 86 proof = 43 % alcohol, an additional 453 tons of CO2 were produced. Wine sold by the liter during the same year, assuming an alcohol content of 10%, yielded 183 tons of CO2. About 6 billion gallons of beer sold, assuming 3.4% alcohol, yielded about 608 million tons of CO2. No attempt has been made to estimate the amount of CO2 produced globally from these or other sources of alcohol. They are added merely to illustrate that there are thousands of significant CO2 sources including thousands of micro breweries and small vineyards that make wine. We also are cognizant of the fact that the United States has been around the alcohol production barn before with the 18th Amendment to the Constitution in 1919 and the 21st Amendment in 1933 repealing the 18th and any issue pertaining to ethanol production is unlikely to be revisited in the foreseeable future.

If global warming is due to CO2 in the atmosphere, the production and use of bio-ethanol as an alternative fuel will also contribute to warming. Further, the argument that "CO2 from annually renewable feedstocks is not generally considered as a net contributor to atmospheric CO2" is rather ludicrous. CO2 in the atmosphere is CO2 in the atmosphere, regardless from whence it came. For those who believe that CO_2 in the atmosphere is a very minor contributor to global warming; it doesn't really matter.

However, ethanol use will decrease dependence on imported oil and contribute to both energy sufficiency and to employment of farmers in the United States and elsewhere. The larger question to be considered is: *Would Americans prefer to spend tax dollars to subsidize Americans or to contribute those dollars via higher oil prices, to foreign oil exporting countries that support and harbor enemy extremists?*

Ethanol produced biologically from the carbohydrate categories mentioned have already released CO_2 in the weight ratio of 22 CO_2 from every 23 ethanol produced, whether in pounds, tons, grams, or other weight measures. Therefore, production of 10 billion gallons of ethanol (density 0.79 at 20 C.) will have already generated 39 million tons of CO_2; production of 5 billion tons of ethanol would result in half that much CO_2 and 20 billion tons of ethanol will result in twice the amount of CO_2. For details on the formation of alcohol plus CO_2 from carbohydrates, readers are referred to any college level text on biochemistry published in the last 40 years.

The bottom line relative to the comparison of ethanol as an alternative to gasoline in a global context is that seed corn or some other starting material is the starting source; not ethanol. All six carbon atoms found in each hexose sugar present in starch, cellulose, the more complex sugars or other hexose sugars when converted to ethanol and then combusted, each become oxidized to CO_2.

Benign and Beneficial Environmental Effects of Increased Atmospheric CO2.

A recent thorough review of the environmental effects of increased atmospheric CO2 pointed out that the Earth's temperature has varied only about 3 degrees C. over the past 3000 years. The coldest period during the past 500 years was only about 1 degree C. below average. This was reached during the Little Ice Age in the early 1700s as determined by surface temperatures in the Sargasso Sea; a large area in the Atlantic Ocean, which was estimated by isotopic ratios of the remains of marine organisms in bottom sediments. During the period 1700 to 2000 the authors analyzed the shortening of glaciers; a process that has lasted about 180 years to date and began well before the rapid increase in use of fossil fuels just prior to World War II. These data also closely corresponded to: changes in solar activity, changes in Arctic air and United States surface temperatures, United States rainfall and tornado frequency, Atlantic hurricanes that made landfall, maximum hurricane wind speed, the number of violent hurricanes, and they were able to discern that sea level increase began about 100 years prior to 1940.

All of the above analyses are discussed by the authors relative to global warming hypotheses and led to the conclusion that world temperature is controlled by natural phenomena; not to the increase in atmospheric CO2. Also discussed are difficulties with interpretations of modeling data often made by computer modeling experts who may not have sufficient expertise in the environmental processing subject matter (Ref. 33). Their figures illustrate that glacier shortening occurred before and is unaffected by the post 1940 increased fossil fuel use; that arctic air temperature correlates with solar activity, not increased fossil fuel use after 1940 and they present a composite of seven independent records over the time period 1950 to 2000 .

There is ample experimental evidence to show that increasing the concentration of CO2 stimulates plant growth. "CO2 Science" (Ref. 2, 3,) occasionally reviews articles on plant growth responses to atmospheric CO2 enrichment that have been published in the peer reviewed scientific literature (Refs.2, 3). The information recently reviewed in the article by Robinson et al.(Ref.33) on the stimulation of plant growth by elevated levels of CO2, points out the increased growth in long lived pine trees and a 40% increase of forests in the United States over the last 50 years. The authors also cite considerable experimental evidence that CO2 fertilization accelerates growth of wheat, orange trees and many other trees and crop plants and that plant diversity as well as productivity increases in the presence of elevated CO2 concentrations.(Refs. 6, 7, 9, 13-20, 22-28, 31-33, 34-36, 39).

This implies that significantly decreasing the CO2 concentration in the atmosphere will result in diminished rates of plant growth, including plant food crops- something desperately needed by many underdeveloped countries. It also implies that photosynthetic activity has been increasing to compensate for the elevated rate of carbon oxidation that has been occurring due to various causes and that any major program to sequester large amounts of CO2 should be analyzed for both positive and negative consequences.

Observations and Opinions

Conflictions and Contradictions in the Global Warming Debate.

While reading and listening to commentaries on Global Warming over the last few years it became evident that some of what was being said was in conflict with subsequent comments, often in the same presentation. This resulted in an otherwise coherent speaker or writer coming across as a "music man" selling trombones, or worse. This prompted the inclusion of this little section pointing out a few of the more obvious conflictions or contradictions encountered:

* Declaring CO_2 as a major "greenhouse gas" because it traps energy from sunlight near the earth's surface and then recommending sunlight trapping devices, e.g. solar panels as an alternative.

* Declaring water vapor and moisture as the most powerful "greenhouse gas" in the atmosphere followed by emphasis on CO_2 as the major "greenhouse gas".

* Advocating the switching to ethanol as a fuel to remedy to the use of "greenhouse gas" producing fossil fuels is a serious contradiction, if CO_2 is what is included in the term "greenhouse gas". The production and combustion of ethanol will produce as much CO_2 as the use of fossil fuels. If those recommending it were aware of this, they have entered the realm of hypocrisy.

* Recommending curtailment of CO_2 production and then emphasizing exercise as a means of weight loss; a procedure that converts excess weight to CO_2 and heat.

* Environmentalists who rail against use of energy because of the " production of CO_2, a major greenhouse gas", and then condemn forest management efforts to prevent forest fires that produce enormous amounts of CO_2 as well as smoke and noxious vapors.

* Recommending the large scale sequestration of CO_2, a known plant growth stimulant, will decrease plant growth including food crops and ethanol feedstocks.

* Global Warming is a global issue but many of the recommended solutions in the popular media are based on regional or local comparative information. One example is the comparison of data on CO_2 collected at Mauna Loa, Hawaii, then extrapolated by others to the global accumulation of CO_2 without considering any contribution of CO_2 from a relatively local volcano. The frequently observed flawed comparison of the amount of CO_2 released by alcohol as an alternative to gasoline as an automotive fuel have been discussed in the ethanol section. Other flawed comparisons relate to lack of data on atmospheric CO_2 concentration and global temperature. In many instances comparative data are just not yet available but never the less lead to non-valid conclusions.

It's A Matter Of Choice

Neither the United States nor the Developed World is prepared to give up or severely cut back on the use of cars, trucks, planes, railroads, elevators, central heating, air conditioning, large and small power tools, motors, fans, computers, food, alcoholic and other carbonated beverages and electricity and in general all the myriad items which require energy expenditure accompanied by the release of CO_2. It is a fact that the United States and other developed countries are aggressively assisting under-developed countries to develop their electrification and other labor saving systems. Thus the United States and other developed countries are in effect promoting greater use of energy and the simultaneous production of atmospheric CO_2 by under-developed countries.

That being the case, we are left with increasing the conservation and efficiency of use of energy resources; none of which will bring the carbon cycle back into the balance of the past.

Some "fractional environmental activists" appear to be campaigning to return the environmental clock back to the mid 1880s or early 1900s relative to wildlife habitats. That is; they seem to want to let the deer and the antelope roam; presumably on someone else's property or on a preserve that someone else pays for. Fractional because they don't go all the way and trade in their cars for horses and give up their hot showers, cell phones, TV's, computers etc. Yet these same activists seem to forget that trees are living species and that humans are also a significant component of the ecosystem.

People choose to use heat for housing, cooking, travel, etc., and they also work to retain the heat via clothing and many other insulation products. To every ones benefit, the likes of: Sadi Carnot, Thomas Edison, Michael Faraday, Henry Ford, Robert Fulton and countless others opened Pandora's Energy Box with their insight and inventions. As a result most people in the developed world live much more comfortably and productively than did the inventors of a former time. World population will continue to live, grow in numbers, produce food, fiber, and inventions that increase labor productivity via the substitution of energy for manual labor. Increased economic productivity; defined as an increase in the output of goods and services with the same or lesser amount of labor is directly dependent upon the use of labor saving inventions. This will continue to increase per capita demands for energy and to promote an increase in total population; all of which are accompanied by an increased release of heat and CO_2 to the environment. If global warming is due in significant measure to human activity, it will continue to get warmer.

In the United States we annually spend in excess of $2.5 trillion to keep the population active, defended, educated, entertained, fed, growing, healthy, housed, moving, productive, and secure; and therefore consuming greater amounts of energy with its inevitable heat release and that is not likely to change significantly.

One can make the argument that the historical success of the economy is attributable to the availability of low cost energy. The corollary to this will be that increased cost of energy will decrease the economic success of the country.

If our primary energy concern resides with our dependence on and cost of oil imported from either greedy or rogue countries bent on doing us harm, several options are available; all of which will continue to contribute to global warming. Although, energy sources alternative to imported oil will not decrease global warming; they will provide a means to decrease dependence on high payments to rogue countries that harbor terrorists. Any costs associated with maintaining our ability to import foreign energy sources should be included as a cost of that alternative energy and viewed as an offset to any higher costs of developing and using alternative sources such as nuclear, ethanol, coal cleaning or gasification, solar, wind, etc.

If concern is primarily about global warming produced by human activity, our focus needs to be on greater conservation and efficiency in the use of energy resources. Curtailing consumption of some energy sources must be done with caution. The 1970s oil embargo taught us that when energy consumption was markedly decreased; utilities and others lost revenue but also had fixed costs, requiring rate increases to maintain and service their infra-structure.

If concern is primarily a decrease of CO_2 formation or removal of existing CO_2, with the belief that CO_2 is a major cause of global warming, our focus should be on processes to chemically or biologically convert the CO_2 to reduced organic materials and to minimize processes that oxidize already reduced carbon. For example: "flaming-off" of methane and other combustible gases, non-necessary degradation of plastics, most of which are petrochemicals and other already reduced waste materials. In retrospect, policy decisions such as the development of bio-degradable plastics in the 1970s, may have been short sighted because most plastics are high energy materials that could have been collected and recycled into synthetic lumber and other useful products. Finally, of course, rapidly extinguishing wildfires and where possible, actively managing forests to prevent fires.

Most discussions about global warming and "greenhouse gases" in the popular media appear to be presented by those who slighted their inquiry into physical chemistry, biochemistry and biological processing and are almost exclusively centered on fuel combustion as the source of CO_2 and global warming. News articles containing the phrase: "*….production and release of CO_2, a major greenhouse gas,….*" sends up a red flag warning to the audience that the author or speaker has a lack of understanding about what is being conveyed. Most of us have used the term "greenhouse gas" at one time or another and it is time to stop using that misleading language.

Avoidance of consideration of measures to control large sources of CO_2 other than the combustion of fossil fuels prompts questions as to why those who still believe CO_2 is a primary cause of warming, fail to consider other reasonable options. In this context it is pointed out that CO_2 gas can be considered fungible in the environment, much in the same way that money is said to be fungible by economists. That is, CO_2 gas enters the environment from a multitude of sources and once it is cycling in the environment, one cannot discern under ordinary circumstances (e.g. unless it happens to be labeled with a radio-isotope) from where it originated or where it will reside. Beyond conservation measures, which are always appropriate, it seems that rapidly controlling wildfires (and other unnecessary oxidative degradation processes) would be like "cherry picking" and would serve the *simultaneous twofold purpose of prevention of CO_2 and heat release by combustion plus removal of CO_2 already in the atmosphere* by photosynthetic sequestration into trees and other plants.

The general population in the industrially developed world wants and indeed demands more energy and they do not take kindly to power outages or brownouts. Neither do they want wildfires and other non-essential combustion. Even those who continue to espouse "greenhouse gas" as a major cause of global warming in the face of mounting contrary data, are not likely to forego cars, trucks, planes, railroads, powered ships, lighting, central heating, hot showers, air conditioning, motors, computers, TV's, etc. Actually, the U.S. is assisting lesser developed countries with designs and knowledge so that they can also substitute energy for manual labor and improve their standard of living.

All of the technical information in this document is known by highly competent physical chemists, biochemists, chemical engineers and process oriented biologists, many of whom are employed by major chemical, energy and other high technology companies as well as in universities and other research organizations..

So what is going on? One possibility resides with "Cap and Trade", the latest scheme to make money from a non-issue issue based on tradable CO_2 credits. The "Cap" is an upper limit on how much carbon emission a country can release; with decisions as to what will be the upper limit being made by international consortia, many members of which are representatives from underdeveloped countries.

Hence, an international "Carbon Market" has been created, whereby a country that produces carbon emissions in excess of the group determined "Cap" can "earn" credits by paying (i.e. trading) for the emissions that the underdeveloped countries are allocated but do not need to use. The main carbon emission in the scheme is CO_2; the byproduct of extracting and using energy from any and all organic substances including fossil fuels, food, wood, organic waste, etc.

Reforestation is one of the ultimate keys to cycling CO_2 back into a reduced energy rich substance (wood, grass, etc.). Some countries are indicating they will assume credits for planting trees and sell the credits to developed countries or by allowing companies from developed countries to provide the work and expense, in trade for the credits. The United States is exploring methods for sequestering CO_2 in the deep subsurface where some of the CO_2 presumably would react under alkaline conditions to produce mineral carbonate (e.g. limestone) deposits over time. One wonders if seashell farming or coral reef restoration or reduction of the 5 million acre per year wildfire norm will qualify.

The bottom line on all of the Cap and Trade manipulation is looking like another scheme to either transfer wealth from developed countries to the undeveloped ones or a justification for either a hidden or transparent Carbon Tax in the United States.

Keep in mind; the reason that CO_2 is deemed undesirable is that it is purported to trap enough solar radiation to generate enough heat to be a major cause of the global warming trend. However, there is little if any attention given to direct heat generation and the whole discussion is becoming circular. In any case, there is no credible hard data to show that the amount of solar radiation absorbed by the amount of CO_2 in the atmosphere generates more heat than that generated by direct combustion plus animal and microbial metabolism plus interaction of solar rays with the collective substances other than CO_2 found in the solar light path. Perhaps "Cap and Trade" incentives should be linked directly to reduction in global temperature rather than diminished CO_2 formation.

It was recently reported that the Chairman of the U.S. House of Representatives Energy and Commerce Committee indicated that "….a cap and trade system alone doesn't convey the real cost of climate change, since it puts its primary cost on companies, which then pass that burden on to consumers via higher prices. It's a hidden tax, …" (Ref. 36).

The whole premise of "Cap and Trade" as applied to CO_2 and reduced carbon substances is, in general, reminiscent of a quote from former President Ronald Reagan referring to the United Nations on a different treaty topic, who said: " No (national) interest of ours could justify handing sovereign control of two thirds of the earth's surface over to the Third World. No one has ruled out the idea of a treaty--one which makes sense--but after long years of fruitless negotiating, it became apparent that the underdeveloped nations who now control the General Assembly were looking for a free ride at our expense--again."

Developed countries underwent development because their populations worked harder (longer hours), smarter (education, inventions), had appropriate incentives (pay and freedom) and in some cases natural resources. The developed countries have and continue to assist under-developed and developing countries; several of which appear to be trying to kill the goose that lays the golden eggs. They often take a short sighted view that the developed countries should transfer wealth directly to them without their first initiating the necessary incentives and habits to allow themselves to develop. It would appear to be a better course for the underdeveloped to promote the developed and keep them developing so that there would be more wealth generated to transfer to the underdeveloped. Many underdeveloped countries also appear to be missing a golden opportunity to test market or promote proof of principle studies to develop alternate energy sources.

Conclusions

* Observations and data analyses indicate that the present Global Warming trend is part of a natural cyclical phenomenon brought about primarily by solar radiation with lesser contributions from hundreds, perhaps thousands, of essential human activities. Therefore, greater emphasis should be devoted to energy security and independence than to any minor effect CO_2 may have on global warming.

* The world is fortunate that CO_2 is only a very minor contributor to global warming both because it is so all pervasive in life processes and because this conclusion supports the scenario that all types of alternate energy, e.g. ethanol, solar panels, nuclear, coal gasification and liquefaction, wind, etc. can be used to promote energy independence even though many alternate fuels will produce CO_2 and none will significantly reduce global warming. It also indicates that those who practice physical exercise to promote their health need not be concerned that they are significantly contributing to global warming.

* There is ample technical evidence available to demonstrate that increased CO_2 levels stimulate plant growth, including trees and crop plants. It is concluded that any large effort to sequester and remove CO_2 from the atmosphere is likely to result in decreased plant growth, including food crops and therefore be a counterproductive waste of money that could be put to better use elsewhere. One such alternative is to increase efforts to manage forests and more rapidly extinguish forest fires.

* All organic substances (i.e. carbon containing substances of biological origin synthesized by organisms), as well as additional substances derived from them such as: petrochemicals, soaps, detergents, dyes, paints and other coatings, many explosives, natural and man made rubber, most fabrics, plastics and innumerable others; whether fiber, food, fuel or waste products, contain carbon that when oxidized by combustion, aerobic respiration or fermentation will release CO_2 to the environment. All photosynthetic life requires CO_2 as a nutrient. To brand CO_2 as an environmental pollutant is the epitome of absurdity! It will taint a large segment of the economy.

* Thermodynamics teaches that the use of all energy produces some heat and non-useable energy called entropy. Therefore, production of heat in the modern world is inevitable.

* Repeated conversion of one energy type into another type will add to the net production of heat and entropy.

* The scheme of "Cap and Trade" appears to be more a ploy for either the transfer of wealth from developed to underdeveloped countries or a possible means for initiating a carbon tax, than as a means to control global warming or promote energy independence.

* Economists assumption that raising the cost of energy will eventually decrease demand for energy does not appear to be entirely valid, particularly in the case of oil or fuel use for several reasons. Most workers must continue to travel to work or lose their wages. Most deliveries of goods and services

via ships, planes and trucks will continue and the added fuel cost will be passed on to the consumer; providing them with a dual economic debit. Most citizens who live in cold regions must continue to heat their homes in winter regardless of costs. Finally, in the case of North America; the distance traveled in the pursuit of commerce is considerably longer than that of many critics in Western Europe, requiring greater energy expenditure.

* Low cost energy contributes significantly to the development and use of labor saving inventions, that in turn increases productivity and contributes to a modern industrialized economy. As a corollary; a significant and sustained increase of the cost of energy will weaken the economy.

* Conservation efforts to reduce consumption of energy are always a valid objective within the constraint that energy suppliers must be able to maintain their infrastructure and service their fixed costs.

* The United States should revive Arbor Day; a non-tax "Green" means for promoting tree planting. It is noted that the Arbor Day Foundation with their Tree Line USA Program honoring 149 U.S. utility companies is doing just that.

On The Lighter Side

Allegedly, the great philosopher, Jimmy Buffet, sang in words and music at a recent U.N. banquet on the environmental impact of methane from cattle: "If we couldn't laugh we would all go insane."

Admiral Heatly, after meeting with the Joint Chiefs, announced they were questioning the recommendation of a United Nations Sub-Committee on Energy Conservation, that they mothball the nuclear fleet in favor of the new and lighter 80 meter polycarbonate sails.

Next time you are out to dinner order a scotch and a splash of pollution. Keep an eye on the bartender!

What do you mean the beer tastes flat? Our new boiling process drove out all that noxious "greenhouse gas" contamination! Keep your eye on the bartender!

Hand a judge a citation for polluting the environment next time you see one open a can of Coke or a bottle of wine. Oops! They may not be liable if the waiter served it. But, you might sue for contaminating you with that horrible gas from the greenhouse, alas.

Order a glass of woodchip wine or a switch grass cocktail. Keep an eye on the bartender!

Oil producing countries have agreed to contribute money proportionate to their output of crude oil that ultimately contributes to atmospheric CO_2.

Countries that export wine and spirits to the U.S. have agreed that since the corresponding amount of CO_2 was produced in their home country they are willing to pay any proportionate "Cap & Trade" levies on alcoholic beverages to the U.S.

References and Citations

1. Burwell, C. C., 1978. *Solar Biomass Energy: An Overview Of U. S. Potential*, Science, Vol. 199, 1041-1048.

2. *CO2 Science Magazine*, a weekly on line magazine published by the Center for the Study of Carbon Dioxide and Global Change, P.O. Box 25697, Tempe, Arizona, U.S., 85285. **URL** http://www.CO2science.org

3. **Volume 5, no.32, August 7, 2002.**

4. **Volume 11, no.4, January 23, 2008.**

5. CRC Handbook of Chemistry and Physics. *An updated Volume Published Each Year by the CRC Press.* Boca Raton, FL., Ann Arbor, Mich., London, Tokyo.

6. Cure, J. D., and B. Acock. (1986) *Agr. Forest Meteorol.*, 8, 127-145.

7. Cyr H. and M. L. Pace. (1993) *Nature*. 361, 148-150.

8. Dessler, Andrew, and Edward Parson. (2006) *The Science and Politics of Global Climate Change, A Guide to the Debate. Cambridge University Press.* Cambridge, New York, Melbourne, Madrid, Cape Town, Singapore, Sao Paulo. 190 pp.

9. Drake, B. G. and P. W. Leadley, (1991) *Plant, Cell and Environment.* 14. 853-860.

10. Flannigan, M. D., J. B. Stocks and B. M. Wotten. (2000) *Climate Change and Forest Fires.* The Science of the Total Environment, 262, 221-229.

11. Garrels, Robert M., Fred T. Mackenzie and Cynthia Hunt. (1975) *Chemical Cycles and the Global Environment; assessing human influences.* 206 pp. William Kaufmann, Inc., Los Altos, California.

12. George, K., L.H. Ziska, J.A.Bunce, and B.Quebedeaux. (2007) *Elevated Atmospheric CO2 Concentration and Temperature Across an Urban-Rural Transect.* Atmospheric Environment, **41,** 7654-7665.

13. Gifford, R. M. (1992) *Advances in Bioclimatology.* 1.24-58.

14. Grace, J., J. Lloyd, J. McIntire, A. C. Miranda, P. Meir, H. S. Miranda, C. Nobre, J. Montcreiff, J. Massheder, Y. Malhi, I. Wright, and J. Gash (1995) *Science*. 270, 778-780.

15. Graybill, D. A. and S. B. Idso (1993) *Global Biogeochemical Cycling.* 7, 81-95.

16. Idso, S. B.,1989. *Carbon Dioxide and Global Change: Earth in Transition.* IBR Press
17. Idso, K. E. and S. B. Idso (1974) Agr. *Forest Meteorology.* 69, 152-203.
18. Idso, S. B. and B. A. Kimball (1991) *Agr. Forest Meteor.* 55, 345-349.
19. Idso, S. B. and B. A. Kimball (1994) *J. Experimental Botany* 45, 1669-1692.
20. Idso, S. B. and B. A. Kimball (1997) *Global Change Biology* 3, 89-96.

21. Kerr, Richard A., (1996) *A New Dawn for Sun-Climate Links.* Science, 171, 1360-1361.

22. Kimball, B.A. (1983) *Agronomy. J.* 75. 775-788.

23. Kimball, B. A., P. J. Pinter Jr., D. J. Hunsaker, G. W. G. Wall, R. L. LaMorte, G. Wechsung, F. Wechsung, and T. Kartschall (1995) *Global Change Biology*, 1, 429-442.

24. Lam, S.H. (2007) *Logarithmic Response and Climate Sensitivity of Atmospheric CO2.* 1- 15, www.princeton.edu/-lam/documents/LamAug07bs.pdf.

25. Lawler, D. W. and R. A. C. Mitchell, (1991) *Plant, Cell and Environment.* 14, 807-818.

26. Lindzen, R.S. (2005) Proc. 34th Int. *Seminar. Nuclear War and Planetary Emergencies.* ed. R. Raigaina. World Scientific Publishing, Singapore, 189-210.

27. McNaughton, S. J., M. Oesterhold, D. A. Frank, and K. J. Williams. (1989) *Nature.* 341, 142-144.

28. Mortensen, I. M.(1987) *Sci. Hort.*, 33, 1-25.

29. NASA, (2007) *The Carbon Cycle.* Internet http://.nasa.gov/Library/Carbon Cycle/printall.php

30. National Interagency Fire Center. (2008) *Wildland Fire Statistics.* http://www.nifc.gov/fire_info/fires_acres.htm

31. Pinter, J. P., B. A. Kimball, R. L. Garcia, G. W. Wall, D. J. Hunsaker, and R. L. LaMorte. (1996) *Carbon Dioxide and Terrestrial Ecosystems* 215-250. Koch and Mooney, Academic Press.

32. Poorter, H. (1993) Vegatatio. 104-105, 77-97.

33. Robinson, Arthur B., Noah E. Robinson and Willie Soon. (2007) Environmental Effects Of Increased Atmospheric Carbon Dioxide. *Jour. American Physicians and Surgeons,* (2007), **12, 79-90.**

34. Scheiner, S. M. and J. M. Rey-Benayas (1994) *Evolutionary Ecology.* 8, 331-347.

35. Smith, W. B., P. D. Miles, J. S. Vissage and S. A. Pugh, (2002) *Forest Resources of the United States,* U. S. Forest Service and U. S. Dept. of Agriculture.

36.. Strassel, Kimberly, (2007) *Some Inconvenient Truths. The Weekend Interview with John Dingell.* Wall Street Journal, October 6-7, A21.

37. United States Department of Energy. (1991) Global Climate Trends and Greenhouse Gas Data. Report to Congress of United States. 185 pp, plus Appendices. NTIS-PR-360.

38. United States Congress, Office of Technology Assessment (1991). *Changing By Degrees: Steps To Reduce Greenhouse Gases,* OTA-O-482. (Washington D.C: U.S. Government Printing Office. February 1991).

39. Waddell, K. L., D. D. Oswald and D. S. Powell, (1987) *Forest Statistics of the United States,* U. S. Forest Service and U. S. Dept. of Agriculture.

Appendix A

A. List of Some Energy Conversion Factors.

Energy and Electricity

In general the following prefixes in the American numbering system are used:

kilo = 1000, mega = 1million, giga = 1 billion, tera = 1 trillion,
1 quad = 1 quadrillion   1 followed by 15 zeros.

1 British Thermal Unit (BTU) = 252.36 calories = 0.252 Cal.

1 BTU = 1055 Joules , 1 BTU = 778 foot pounds of work

1 QUAD = 1 Quadrillion BTUs = 10 X 15th power BTUs

1 calorie is the amount of heat energy required to raise the temperature of 1 gram of water by 1 degree Celsius (C.), also sometimes called 1 degree centigrade.

1 calorie = 4.184 Joules

1 degree C. = 1 degree Fahrenheit -32, divided by 1.8.

1 degree Fahrenheit (F.) times 1.8, + 32 = 1 degree C.

1 degree Kelvin (K.) = 1 degree C.-273. 0 degree K. is the absolute temperature.

1 HP (horsepower) motor = 746 watts of power = 2545 BTUs per hour

1 horse power = 2545 BTUs per hr.

1 watt = 1 ampere at 1 volt = 1 joule per second.   1 KW = 1000 joules per second

1 KWH (kilowatt hour) = 1 KW work for 1 Hr. = 3412 BTUs

1 cubic foot of natural gas = 1000 BTUs (approx.)

1, 31.5 gallon barrel of Jet Fuel = 5,670,000 BTUs (approx.)

1, 42 .0 gallon barrel of crude oil = 6000 cu. ft. of natural gas (approx.)

1 short ton (2000 lb.) bituminous coal = 24,000,000 BTUs (approx.). 1 lb. of average coal = about 13,000 BTUs.

1 short ton x 0.907185 = 1 metric ton

1 short ton = 7.5 (31.5) gallon barrels.

1 31.5 gallon barrel gasoline = 5,248,000 BTU

1 42 gallon barrel crude oil = 5,800,000 BTU

1 short ton of electricity generating grade coal = 22,400,000 BTUs (approx.).

BTUs from 1 short ton coal = BTUs from 4.17 barrels crude oil = BTUs from 25,000 cu.   ft. natural gas. Approximately.

In 2001 in U.S., 3.43 trillion KWH of electricity were sold . (29% from non-CO_2 generating sources and 71% from fossil fuels).

Oil demand was 19.7 million barrels per day (mbd), 8.6 mbd for automobile use.

Natural gas was 2.6 trillion cu. ft.

In 2002 world oil demand was 76.6 mbd.

Volume and Weight Conversion

1 ounce = 28.3495 grams., 1gram = 0.03527 ounce

1 U.S. gallon (gal.) = 3.7854 liters (L.)

1 pound (lb.) U.S. = 4536 gram (gm.) approx. = 0.4536 kilogram (kg.)

1 kg. = 2.2 lb. approx.

1 short ton = 2000 lb., 1 metric ton = 0.907 short ton approx.

1 metric ton = 1.10 short tons

1 metric ton = 2,204.6 lb.

1 metric ton = 1 million gm.

1 million metric tons carbon = 1 petagram C = 1 PgC.

1 CO_2 weight x 0.27 = 1 carbon weight., 1 carbon weight x 3.667 = 1 CO_2 weight.

1 acre = 0.4 hectare.

Length and Wavelength Conversions

1 inch = 2.54 centimeter., 1 cm. = 0.3937 in.

1 meter = 100 centimeters = 1000 millimeters = 1,000,000,000 nanometers

Appendix Continued

B. An Overview of the Laws of Thermodynamics

The *1st Law of thermodynamics* is the law of conservation of energy and states that the sum of all energies in an isolated system is a constant. That is, energy may be changed from one form to another, but it can't be created or destroyed. It was modified to accommodate nuclear fission and considers matter as a form of energy. It is now the law of conservation of energy and mass and is a product of human experience consistent with Einstein's equation: $E = mc^2$; where E is the amount of energy produced by the consumption of a mass of matter, m, and c is the velocity of light.

If m is in grams and c is in centimeters per second, E will be in ergs. The velocity of light is $3.0 \times 10 \times 10$ cm./sec.

This law forbids the existence of a perpetual motion machine that would produce more energy (work) than was used in its operation. "Something From Nothing Ever Comes."

The *2nd law of thermodynamics* predicts the spontaneous direction of energy transformation. It states that all systems change to decrease their capacity for change, i.e. to approach equilibrium. Examples are: water runs downhill, heat flows spontaneously only from a higher temperature to a lower temperature without the expenditure of work on the system, gases spontaneously expand from a higher pressure against a lower pressure, electric charge moves from a higher potential to a lower potential, living organisms at a point in time spontaneously begin to undergo ageing. During spontaneous changes the internal energy of a system tends to be progressively less available for doing work. That is, the system increases in *entropy*; a measure of the energy of a system that is unavailable for doing work and which is beyond the scope of this discussion. In effect the 2nd law states that the universe must be running down through natural spontaneous processes that make its energy increasingly unavailable for useful work.

The principle of zero entropy of a perfect crystalline substance at absolute zero temperature is called the *3rd Law of Thermodynamics*. It is a tenet of physical chemists that all atomic motion ceases at absolute 0 temperature; which is, 0 degree K = -273 degree C. or -459.4 degree F. It is also an accepted tenet that absolute 0 K is not attainable because there is no sink at a lower temperature to which the last trace of heat energy may flow. A concept of free energy function; also called net work or isothermally useful work was introduce independently by Gibbs and Helmholtz, and provides a workable criterion for predicting whether a chemical reaction will proceed spontaneously. Calculations of *Gibbs Free Energy* for most chemicals are now available in handbooks of chemistry and physics for use in determining whether a reaction will proceed spontaneously and how much free energy is either required or evolved under selected conditions at specified temperatures.

C. Anecdotal caveat. On August 27, 1883 the volcanic Maylasian Island, Krakatoa exploded completely eliminating the island, dissipating the equivalent of 6 to 8 cubic miles of earth into the atmosphere as a suspension of fine colloidal particulates. The particulates were reported to have encompassed the entire globe over the next several years, thereby blocking the sun and altering sunsets which were recorded by artists. The global temperature was also extensively reported to have been lowered by about 1 degree F. but no adequate records were kept of actual temperature lowering. Interestingly, this time period coincides with a leveling off of the temperature as it increased near the end of the Little Ice Age. This could raise questions about the 1 degree global temperature rise over the last 100 to 120 years because the comparative temperature at the beginning of the time span would normally have been about 1 degree higher.

D. The average adult human has a vital lung capacity of about 4.0 L. and at rest exhales about 1 liter (L.) of air 15 times per minute. Exhaled air, called tidal air, contains an average increase of approx. 4.1 volume percent of CO_2 compared to inhaled air at 0.04% CO_2. This calculates out to about 69,000 grams of CO_2 exhaled per year per person or approximately 1.27 Pg. carbon per year by the 6.7 billion world population.

The amount of heat produced by an average, 160 lb. adult person at rest, such as when sleeping or watching a movie, is about 400 BTUs per hour. Although temperatures for warm blooded animals are known (e.g. horses about 100.5 F., cattle and calves about 101.5 F., hogs about 102. F. and sheep at 103 F.) there is no reasonable estimate of the numbers of animals worldwide, therefore there are no reasonable estimates of the amount of heat and CO2 produced, but the collective amount undoubtedly exceeds that for humans.

E. The weight of forest biomass is an unknown but can be approximated by comparing to a known weight of a crop harvest. In 2001 the U. S. sugar cane harvest was 35.1 tons per acre for the 1.027 million acres harvested. Forests are estimated to have about 20 times the biomass of croplands due to their accumulation over several years, with a calculated average biomass accrual of 700 tons per acre. Using the 5 million average number of forest acres burned in the U. S. during the period 2000 to 2006, those fires consumed an approximated 3.5 billion tons of biomass per year. The biomass consists of about 70% moisture (2.45 billion tons) and 30% solids (1.05 billion tons) of which about 60% is carbon (0.63 billion tons). Carbon x 3.667 = 2.3 billion tons CO2 produced per year. If it is assumed that about half of the biomass is burned during a fire, the above values should be divided by 2. Therefore, 1.22 moisture, 1.15 CO2 produced and 0.95 sequestration capacity lost (see below and Ref. 1).

The calculated loss in forest CO2 sequestering capacity is: 20 cu. ft. per acre per year, at 44 lb per cu. Ft. (hardwood) and 32 lb. per cu. Ft. (softwood) or average 38 lb. per cu. ft. = 760 lbs. CO2 per acre per year lost., x 5 million acres = 1.9 billion tons of CO2 sequestration capacity lost per year in the U.S

About the Author

Patrick R. Dugan is an Emeritus Professor of Microbiology at The Ohio State University in Columbus, Ohio and is also a retired Science Fellow from The Idaho National Laboratory in Idaho Falls, Idaho. Dugan holds a B.Sc. degree in Arts and Science, an M. Sc. and Ph.D. degree in Microbiology from Syracuse University. He spent eight years as an Associate Research scientist at the Syracuse University Research Corp., conducting research sponsored by both government and U.S. corporations related to environmental problems as well as food handling, processing, packaging and preservation.

In 1964 he joined the faculty of the Department of Microbiology at The Ohio State University; where he taught and conducted research on various aspects of aquatic contamination by chemicals and microbes and processes to ameliorate aquatic contamination. During this period Dugan served as a Department Chairman and later as the Dean of the College of Biological Sciences. He also served as a Trustee of the Columbus Zoological Association and Zoo and was president of the Ohio Chapter of the American Society for Microbiology from 1968 to 1970. He remains a member of the American Academy for Microbiology.

In 1987 he retired from OSU and joined the Idaho National Laboratory (formerly The Idaho National Engineering and Environmental Laboratory) where he became a Science Fellow and Director of the Center for Bio-processing Technology, conducting research on a variety of projects of interest to the United States Department of Energy and other U.S. agencies while continuing to be a graduate faculty member of The University of Idaho and The Ohio State University. In 1990-1991 he was an Association of Western Universities Distinguished Lecturer. Upon retirement in 1994 he continued as a consultant to the INL.

Professor Dugan's research covered a wide range of disciplines including: sanitation and water pollution; algal, fungal and bacterial control methods and agents; analytical chemistry including methods development; structure and function of exocellular polymers and capsules, chemosynthetic autotrophic bacteria, physiology of aquatic microbes, acidic mine drainage; sulfur and mineral metabolism of microbes; microbial hydrocarbon oxidation and microbial methane formation. His published work includes over 150 articles in peer reviewed journals as well as several book chapters, reports and the book "Biochemical Ecology of Water Pollution," published in both English and Japanese versions. He is listed in Who's Who: in America, in the World, and in several others.

www.ingramcontent.com/pod-product-compliance
Lightning Source LLC
Chambersburg PA
CBHW051101180526
45172CB00002B/728